PROJECT M²: MENTORING YOUNG MATHEMATICIANS

Designing a Shape Gallery

Geometry with the Meerkats

AF836303

M. Katherine Gavin, *University of Connecticut*
Tutita M. Casa, *University of Connecticut*
Suzanne H. Chapin, *Boston University*
Linda Jensen Sheffield, *Northern Kentucky University*

Student Mathematician's Journal

Connections to the Common Core State Standards

Kendall Hunt
publishing company

This material is based upon work supported by the National Science Foundation under Grant No. 0733189 and developed at the University of Connecticut. Any opinions, findings, and conclusions or recommendations expressed in this material are those of the author(s) and do not necessarily reflect the views of the National Science Foundation.

University of Connecticut
Neag School of Education

Kendall Hunt
publishing company

www.kendallhunt.com
Send all inquiries to:
4050 Westmark Drive
Dubuque, IA 52004-1840
1-800-542-6657

ISBN 978-1-5249-1213-0

Published in the United States of America

17 16 15 14 13

Production Date: 2017
Printed by: LSI
United States of America
Batch number: 441213

Table of Contents

Table of Contents

As **Speakers** we will:

1. Talk loud enough for others to hear.

2. Turn to talk to the class.

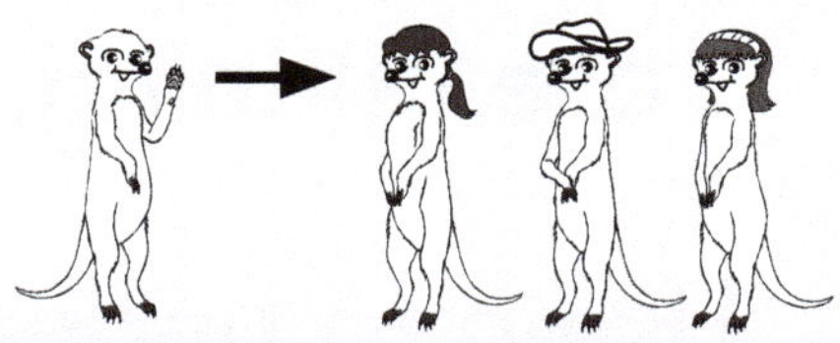

3. Share different ideas.

4. Explain our ideas.

5. Agree and disagree with ideas, not with each other.

As **Listeners** we will:

1. Ask speakers to speak up.
2. Show speakers we are listening.

3. Listen to understand.
4. Ask questions to make sense of the idea.

   Did you mean...?

5. Think carefully about all speakers' ideas.

As **Writers** we will:

Think — Think about the question.

Talk — Talk about the answer.

Tell — Tell:

1. all ideas
2. the answer
3. "why"

Thinking Like a Mathematician*

Here is a list of skills mathematicians use every day. See how many you can use in your Student Mathematician's Journal.

1. Work hard to understand and solve problems.
2. Make sense of numbers in problems and use operations with numbers.
3. Defend your math ideas. Explain why you agree or disagree with someone's thinking.
4. Use the math you know to help solve problems in your world.
5. Choose and use the right math tools to help solve problems.
6. Use correct math vocabulary and symbols to explain clearly.
7. Look for and use patterns to help solve problems.
8. Notice if you are using the same math again and again and look for short cuts.
9. Solve a problem in a new way. Ask new questions to investigate.**

*Adapted from the Common Core State Standards: Standards for Mathematical Practice

National Governors Association Center for Best Practices (NGA Center), Council of Chief State School Officers (CCSSO). (2010). *Common core state standards for mathematics*. Washington, DC: Authors. Retrieved from http://www.corestandards.org/the-standards.

** Johnsen, S. K., & Sheffield, L. J. (Eds.). (2013). *Using the common core state standards for mathematics with gifted and advanced learners*. Waco, TX : Prufrock Press.

Student Mathematician: ______________________________ Date: ____________

THINK DEEPLY

Here are the shapes we noticed in our travels. Some student mathematicians think one group has <u>only</u> triangles and the other group has <u>no</u> triangles.

a. Which group has only triangles?

b. Why are all of the shapes in this group triangles?

Group A

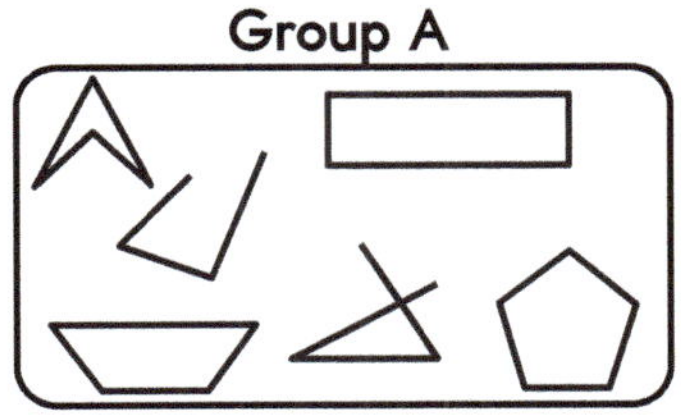

Group B

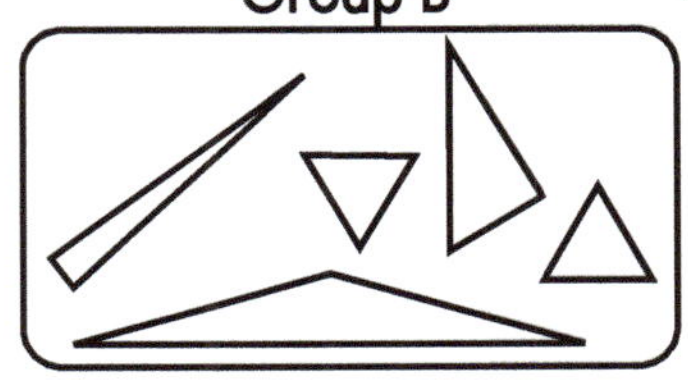

Dear Dru and Teller,

We think that all of the shapes in Group ________ are triangles because ______________________________

and ______________________________

______________________________.

Triangles don't all look the same because ______________________________

______________________________.

and ______________________________. They still are triangles. The pictures of different triangles on the back show what we mean.

Let us know what you think. We can't wait to hear back from you!

______________________________,

____

Student Mathematician: Date:

What Do You See?

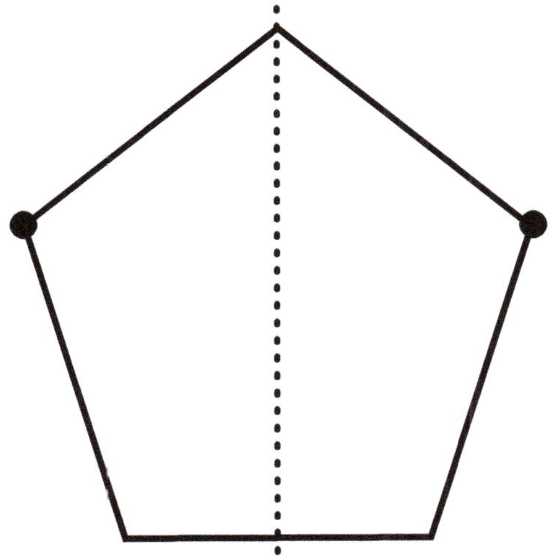

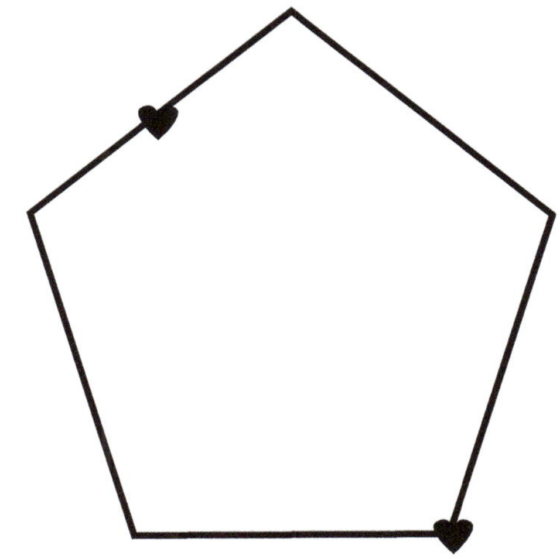

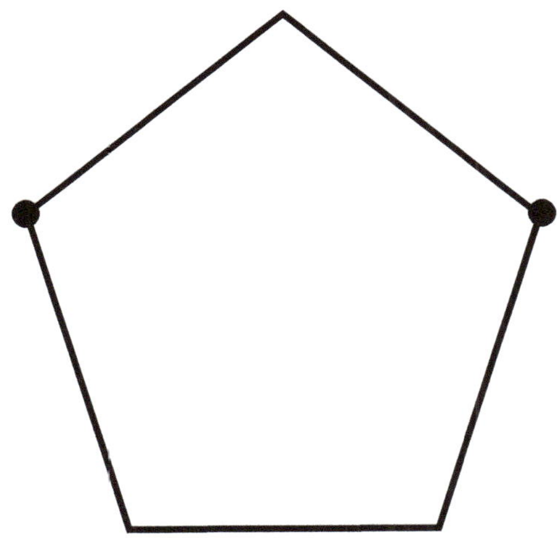

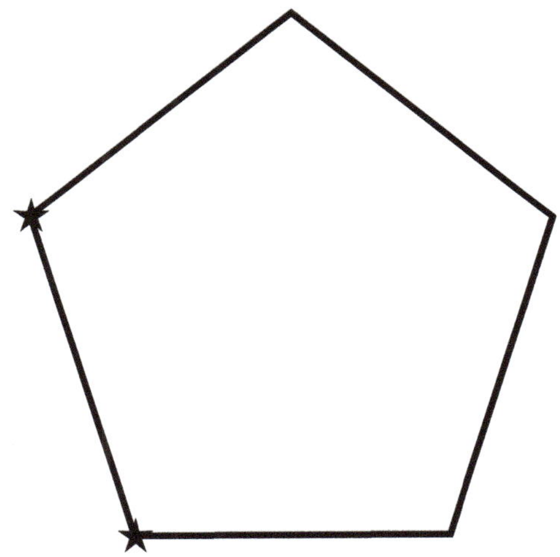

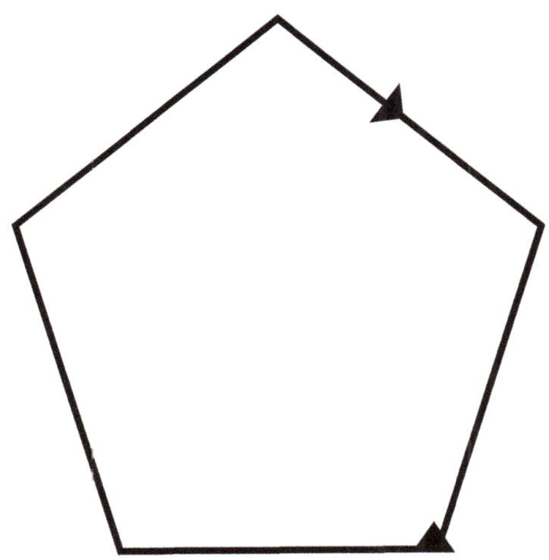

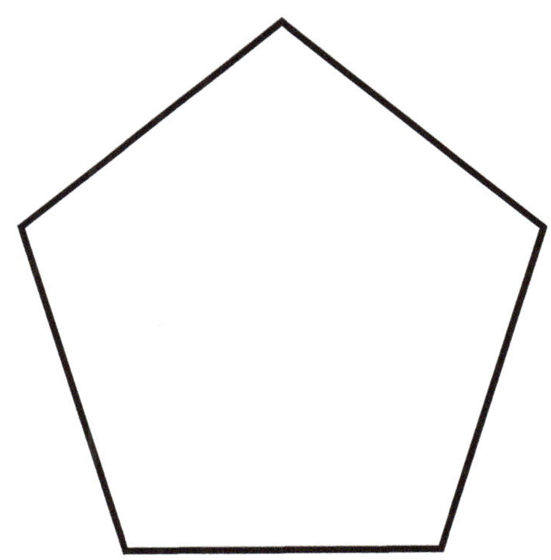

Student Mathematician: Date:

Are These Lines of Symmetry?

Circle "yes" or "no" under each picture. Use a mirror or fold each shape to help you.

1.

Yes No

2.

Yes No

3.
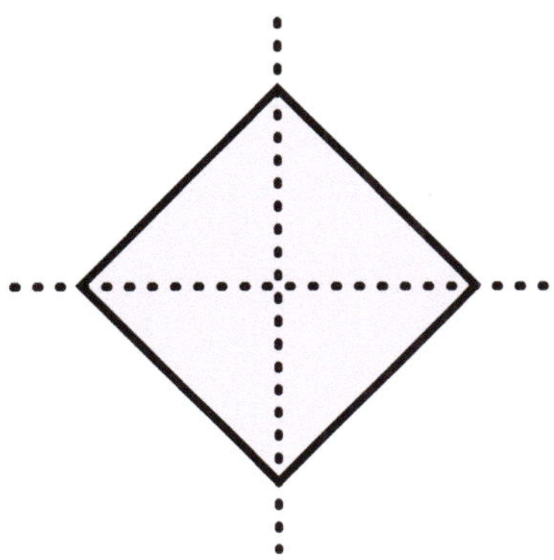
Yes No

4.
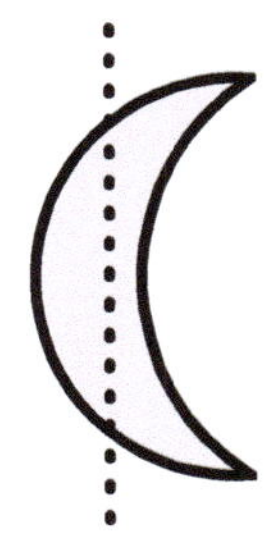
Yes No

5.

Yes No

6.
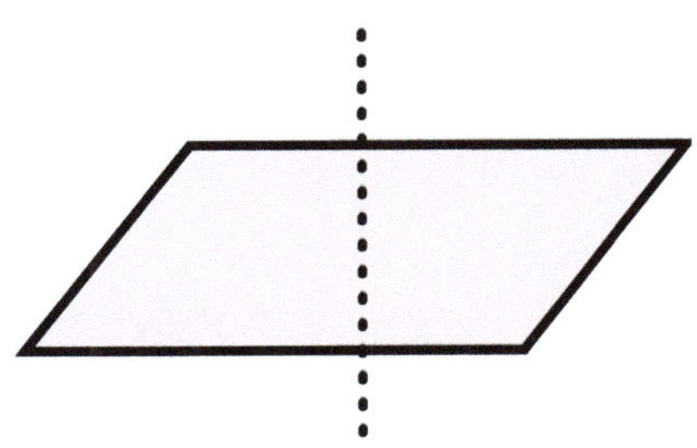
Yes No

7.

Yes No

8.
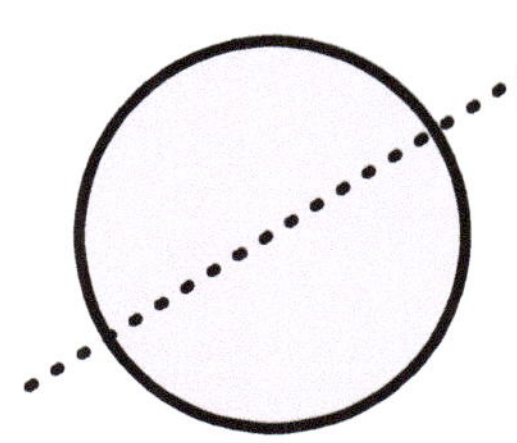
Yes No

Student Mathematician: Date:

Where Are the Lines of Symmetry?

Draw all of the lines of symmetry in each shape below. Use a mirror or fold each shape to help you.

1.

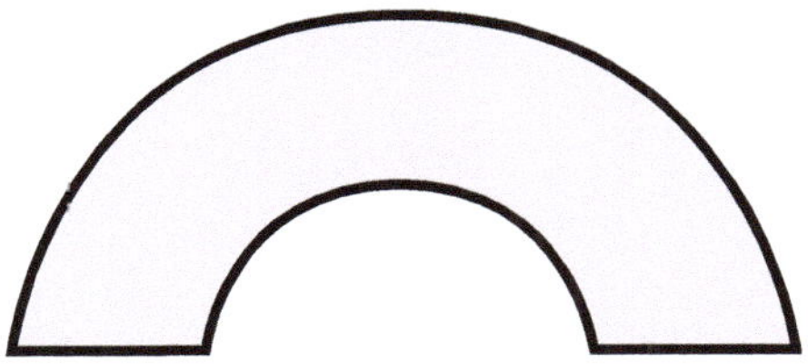

2.

3.

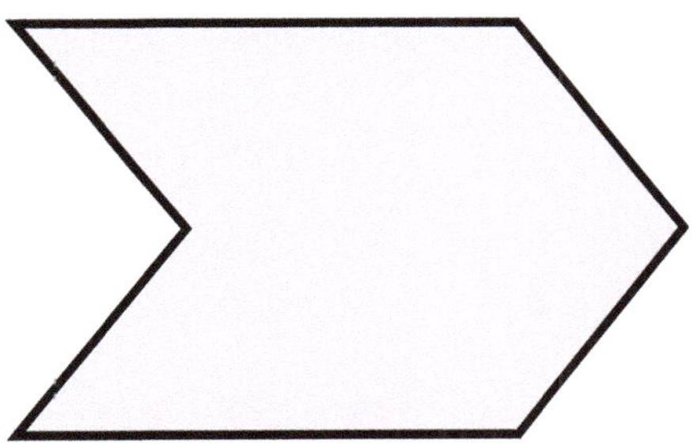

4.

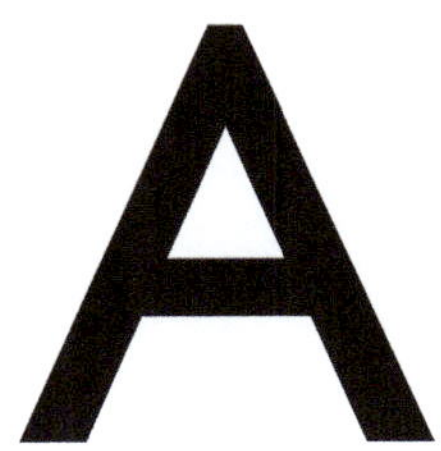

5.

6.

Student Mathematician: Date:

Triangle Designs

Use light-colored and dark-colored paper triangles with your partner to make these designs!

A

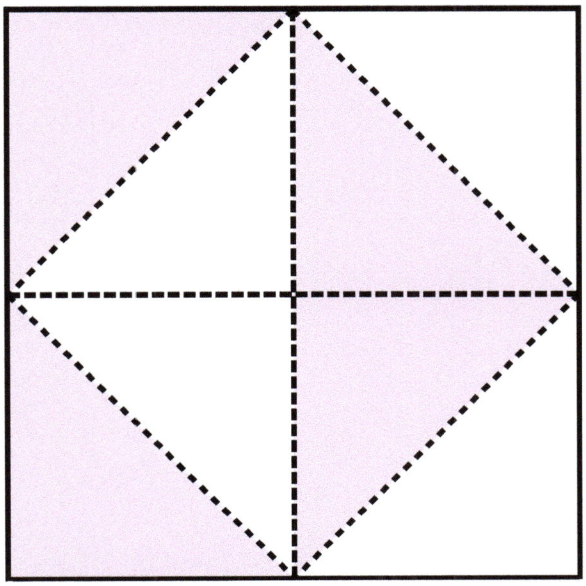

B

C

D

Student Mathematician: Date:

Make the Triangle Designs

A

B

C

D

Student Mathematician: Date:

Triangle Combinations – A

Choose two triangles that have the same letter.
Put them together to make the new shape.

a. Make a new triangle. Trace the new shape. Label each triangle with the correct letter.

Try this more than one way!

b. Make a rectangle that is not a square. Trace the new shape. Label each triangle with the correct letter.

Try this more than one way!

Student Mathematician: Date:

Triangle Combinations — B

Choose two triangles that have the same letter. Put them together to make the new shape.

a. Make a square. Trace the new shape. Label each triangle with the correct letter.

Try this more than one way!

b. Make a parallelogram. Trace the new shape. Label each triangle with the correct letter.

Try this more than one way!

Student Mathematician: ______________________ Date: __________

a. Choose two triangles and combine them to make a rectangle. Trace the rectangle below.

b. How many lines of symmetry does your rectangle have? Draw them using different colors. Tell Dru and Teller how you know these are lines of symmetry.

Student Mathematician: ______________________ Date: ____________

The Transformers – A

What transformation moves each figure from the left to the right? Write flip, slide, or turn on each line.

1. ________________

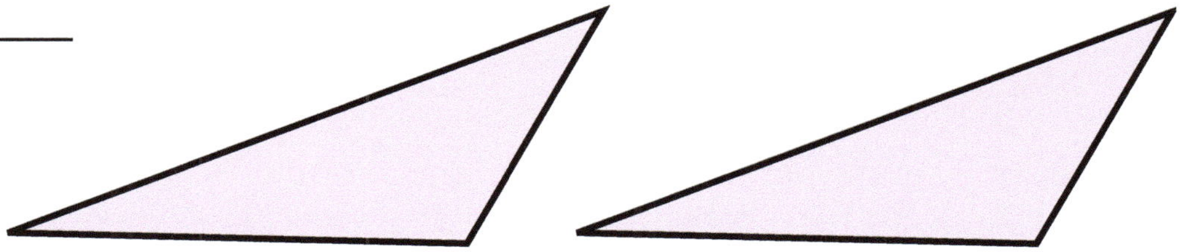

2. ________________

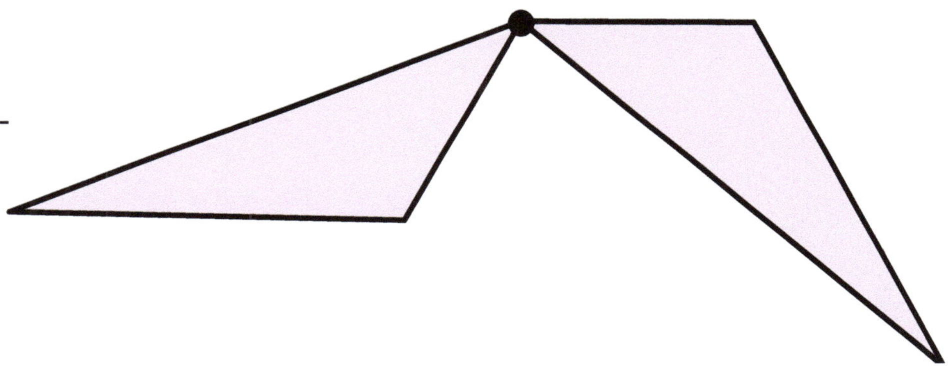

3. ________________

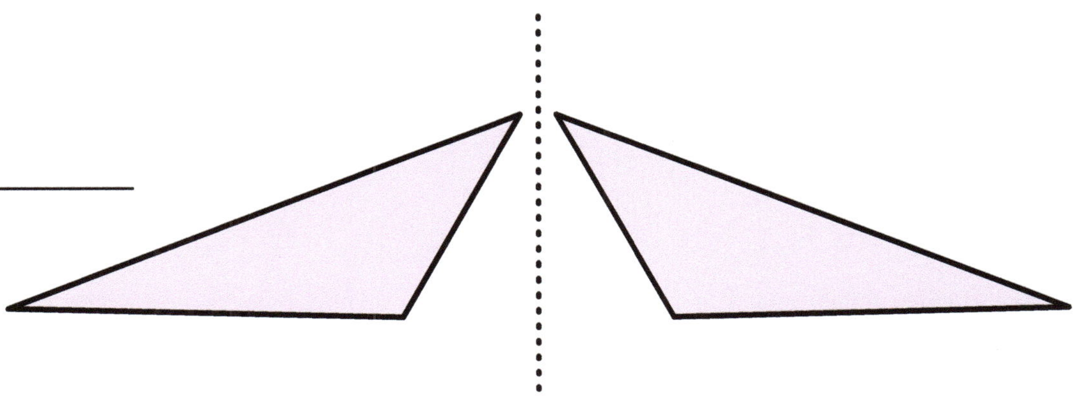

Student Mathematician: Date:

The Transformers — B

What transformation moves each figure from the left to the right? Write flip, slide, or turn on each line.

1. ____________

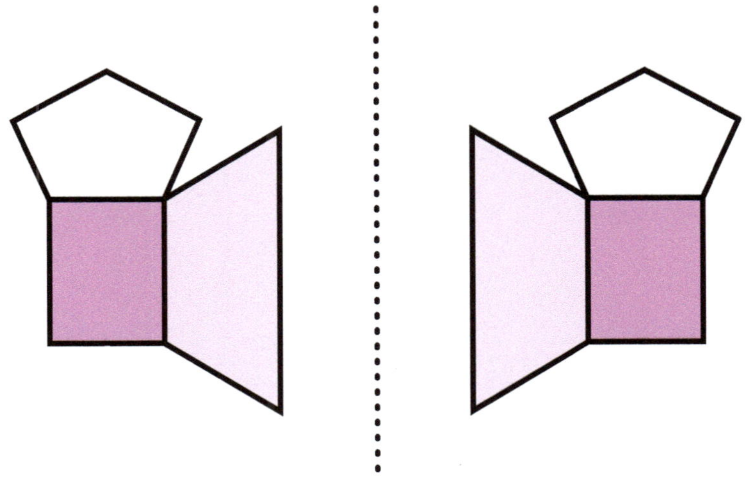

2. ____________

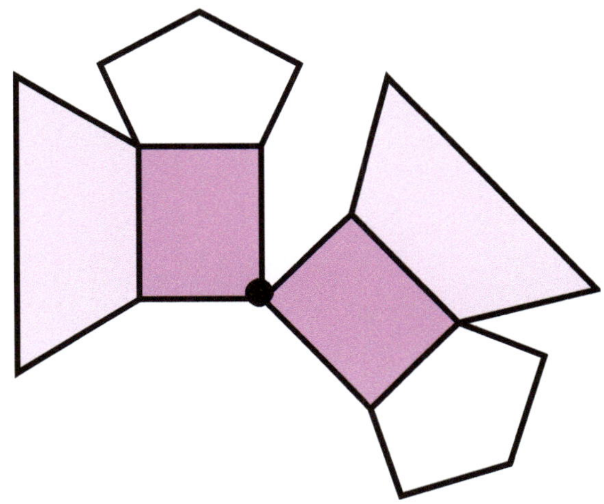

3. ____________

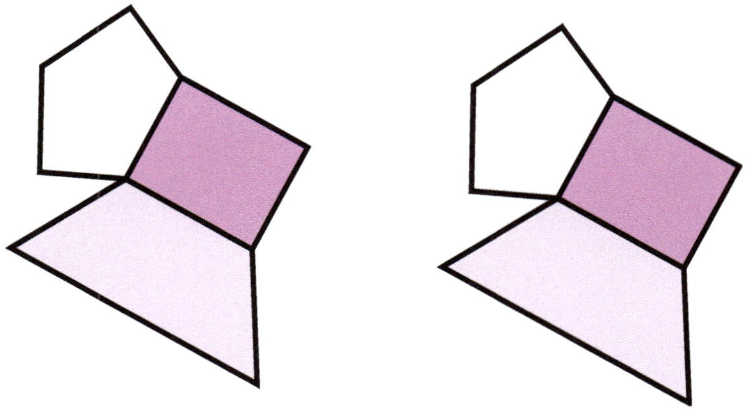

Student Mathematician: Date:

The Transformers – C

Draw each shape after the transformation.

1. flip

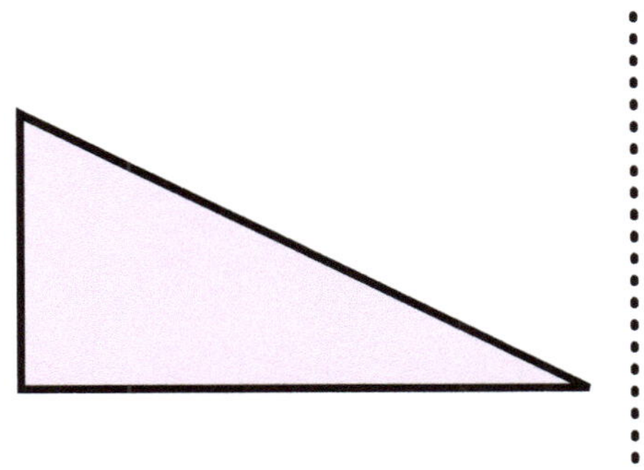

2. slide

3. turn

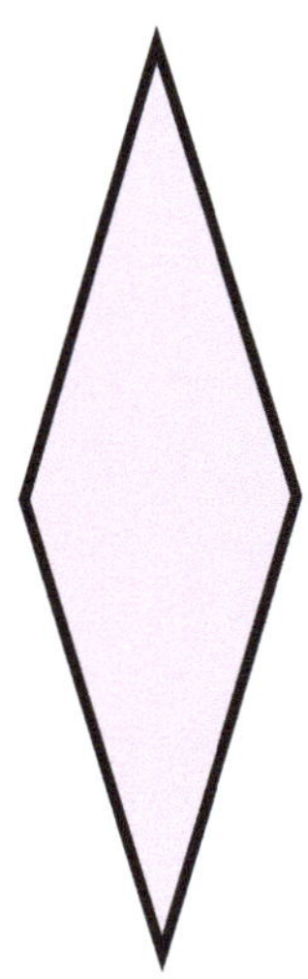

Student Mathematician: ____________________ Date: __________

Transformation Trip

Move the triangle in order. Color each triangle to record the transformation.

Slide = Red
Turn = Blue
Flip = Green

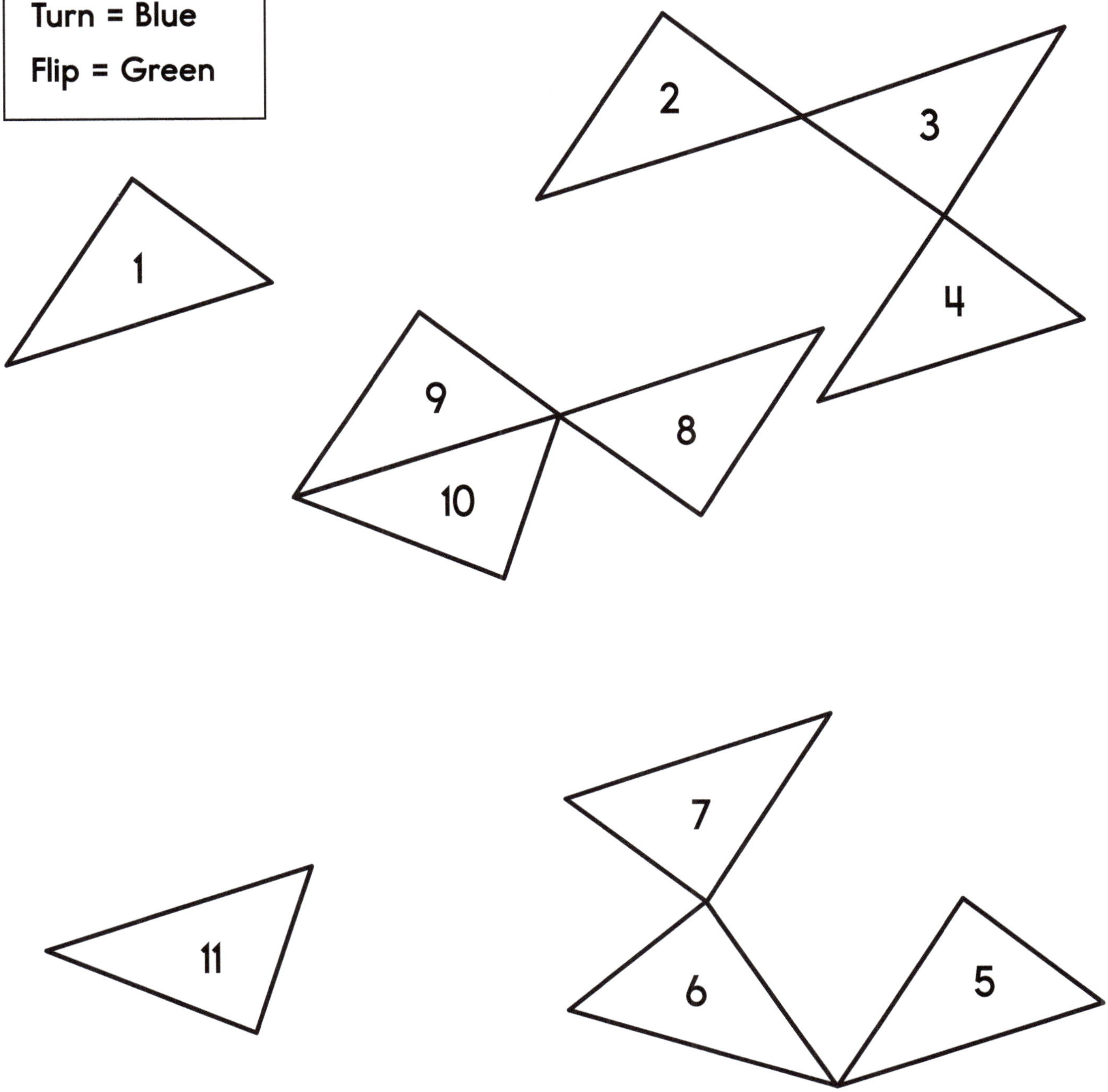

Student Mathematician: ________________________ Date: ____________

Triangle Grid

Shape Gallery
at SMILE
The Sensational Math Institute for Learning and Exploration

Needed:
Scale Models of Your Wall Murals

Send us your designs!

Guidelines

1. Pick 2 or 3 different pattern blocks. Do not use a square or white parallelogram.
2. Make a basic pattern.
   - **a.** Use your 2 or 3 blocks.
   - **b.** Match up at least one side of each block with the side of another block.
3. Trace, color, and cut around your basic pattern.
4. Put your basic pattern on the triangle grid.
5. Lift up your pattern and color your design on the Triangle Grid.
6. Move your basic pattern to cover the Triangle Grid. Color as you go.

Student Mathematician: ______________________ Date: __________

THINK DEEPLY

We need help! Students sent us this scale model of a wall mural. Please tell us:

- The shapes they used to create the basic unit
- How they put the basic unit together
- How they moved the basic unit to create the wall mural

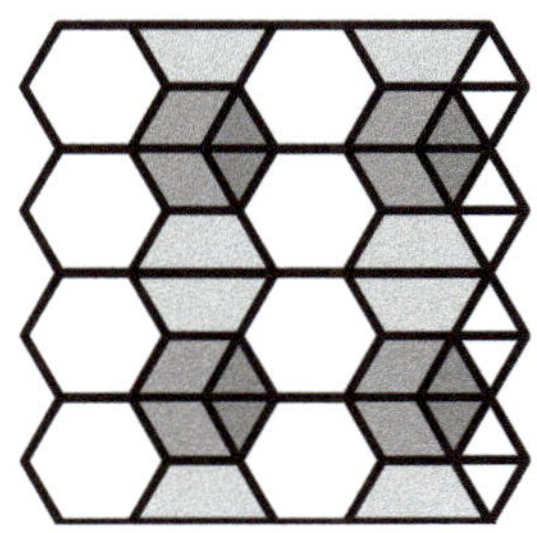

Student Mathematician: Date:

Building 2-D Shapes

Use straws and pipe cleaners to build some of the following shapes.

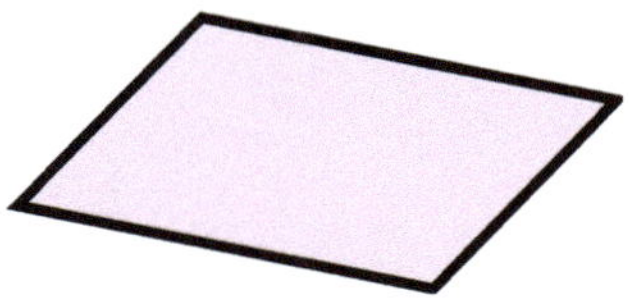
parallelogram

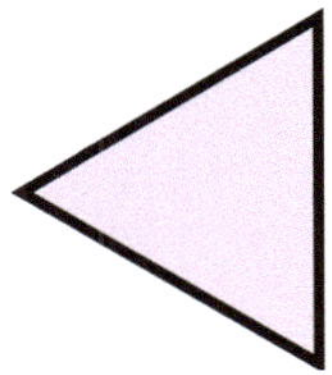
triangle

square

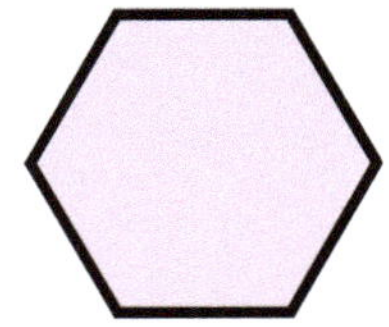
hexagon

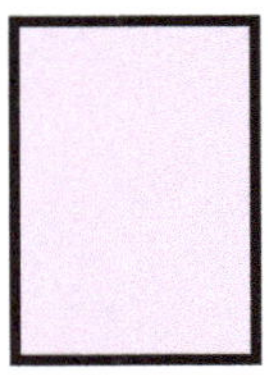
rectangle that is not a square

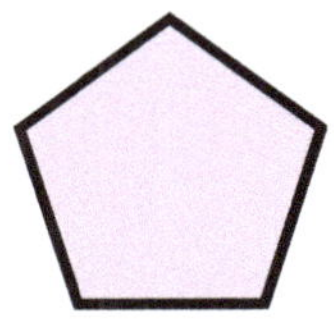
pentagon

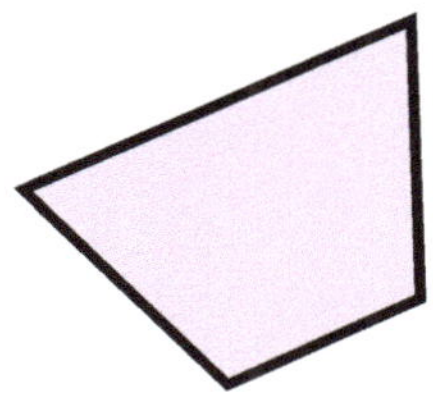
trapezoid

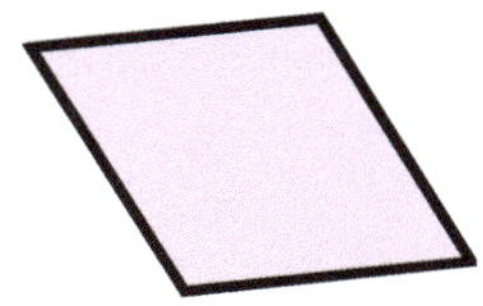
rhombus that is not a square

Shape Name	Describe the Shape

Student Mathematician: Date:

Building Prisms

Cube

You need: 12 short straws

24 pipe cleaners

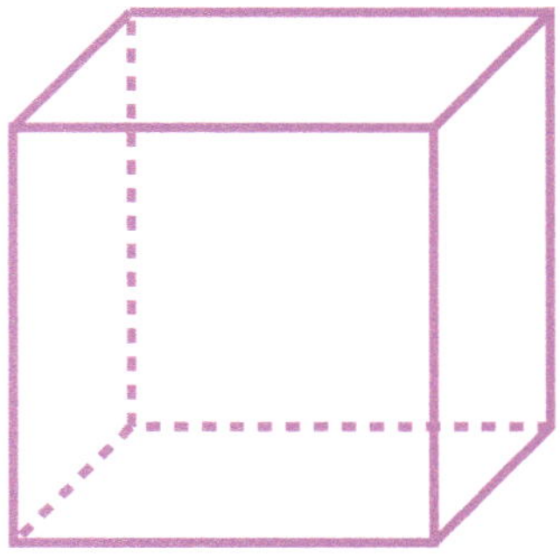

Rectangular Prism

You need: 8 short straws

4 long straws

24 pipe cleaners

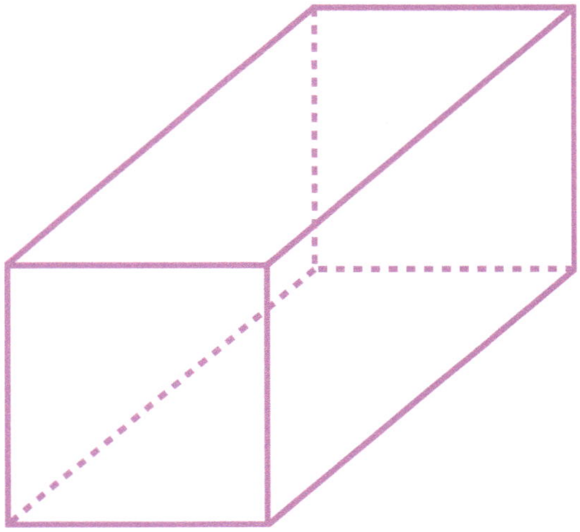

Triangular Prism

You need: 6 short straws

3 long straws

18 pipe cleaners

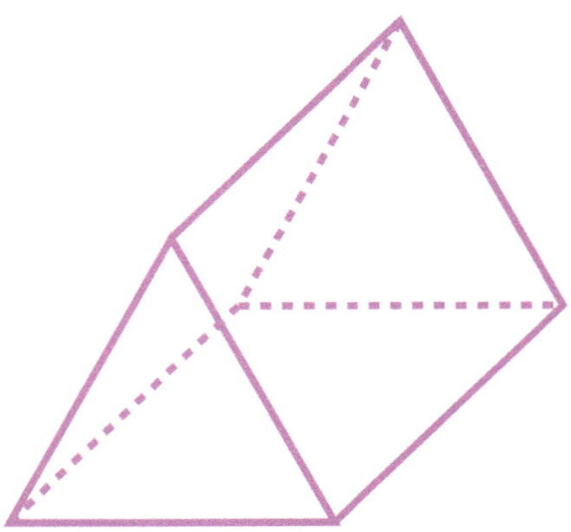

Student Mathematician: ______________________ Date: __________

Building Pyramids

Square Pyramid

You need: 8 short straws

15 pipe cleaners

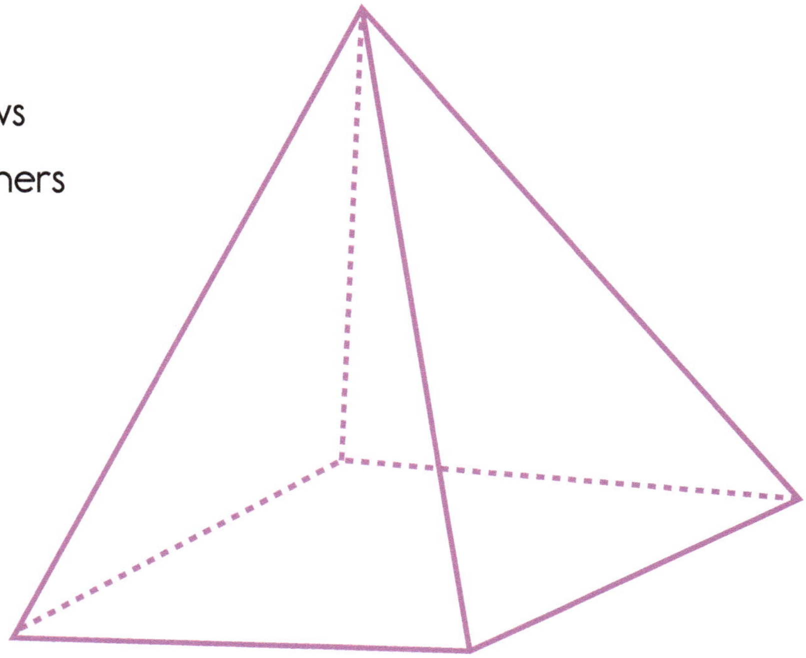

Triangular Pyramid

You need: 6 short straws

12 pipe cleaners

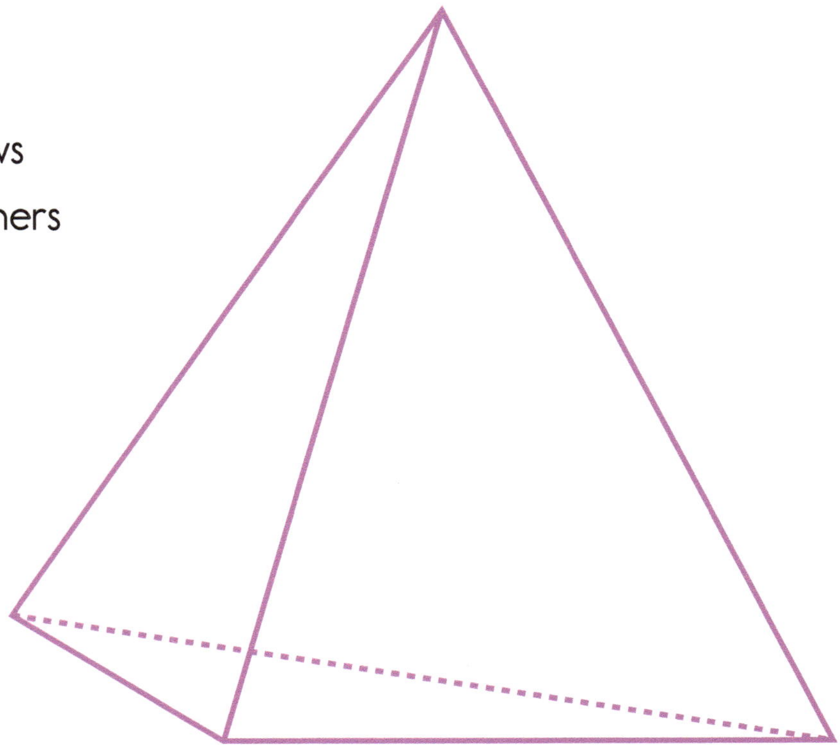

Student Mathematician: Date:

Bubble Experiments

3-D Shape	Shape(s) You Predict Will Be Inside	Shape(s) You Saw Inside

Student Mathematician: ______________________ Date: __________

THINK DEEPLY

We dipped the cube in the bubble solution.

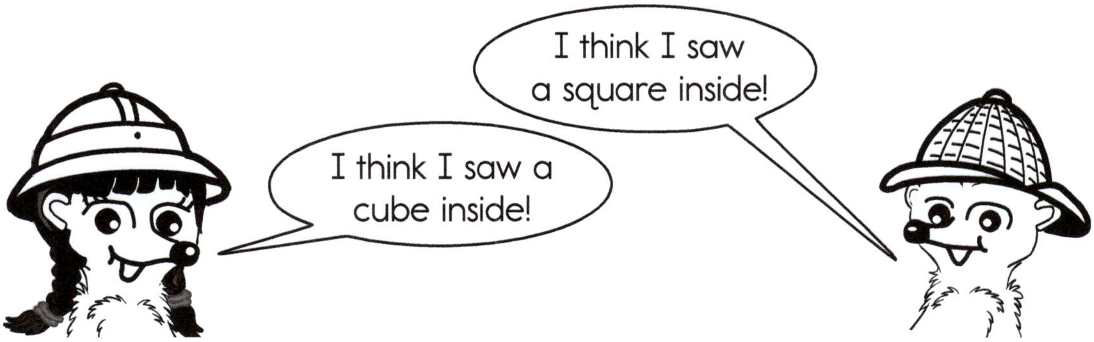

Help us tell the difference between a cube and a square.

Student Mathematician: Date:

Identifying Tangram Blocks

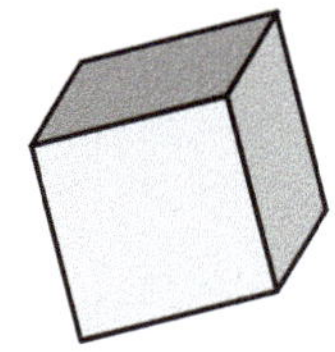

Complete the table using your set of tangram blocks.

Shape	Color	Number and Type of Faces	Number of Edges	Number of Vertices
Cube	Orange	6 squares	12	8
Large Triangular Prism	Red			
Large Triangular Prism	Blue			
Medium Triangular Prism	Yellow			
Small Triangular Prism	Red			
Small Triangular Prism	Blue			
Parallelepiped	Green			

Student Mathematician: ______________________ Date: ______________

Tracing Faces

Name of Block ______________________

Directions

1. "Stamp" a face of the block below by tracing around the face.
2. Lift up the block.
3. Stick an adhesive dot on the face of the block you just traced.
4. On your tracing, make an X in the middle of each shape using a crayon that is the same color as the block.
5. Repeat steps 1–4 until you have traced all faces.

Share Your Ideas

Shape Gallery
at SMILE

The Sensational Math Institute for Learning and Exploration

Needed:
Scale Models of Your Floor Designs

Guidelines

Show us what faces from your blocks will touch the Shape Gallery floor. On your Floor Design Grid:

1. Show a total of four shapes.
2. At least one face should be a triangle. Use a different size triangle if you use more.
3. At least one face should be a rectangle. Use a different size rectangle if you use more.
4. Use four **different** blocks.
5. Make sure the shapes are not touching.
6. Leave at least 30 square inches of the grid uncovered for walking space.

Send us your designs!

Student Mathematician: Date:

Floor Design Checklist

Check your floor design by answering each question. After you have answered all the questions with "yes," ask another pair to check your questions and circle "yes" or "no." If you disagree, talk about it and decide who is correct. Correct your design if necessary.

Question	You		Other Pair	
1. Are there a total of four shapes?	Yes	No	Yes	No
2. Is at least one shape a triangle?	Yes	No	Yes	No
3. If you use more than one triangle, are the triangles different sizes?	Yes	No	Yes	No
4. Is at least one shape a rectangle?	Yes	No	Yes	No
5. If you use more than one rectangle, are the rectangles different sizes?	Yes	No	Yes	No
6. Are the four shapes from four different blocks?	Yes	No	Yes	No
7. Are the shapes not touching?	Yes	No	Yes	No
8. Are there at least 30 square inches of the grid that are not covered?	Yes	No	Yes	No

THINK DEEPLY

Shape Gallery

at SMILE

The **S**ensational **M**ath **I**nstitute for **L**earning and **E**xploration

Help! We want to use a block with a square face for a floor design. Which ones can we use?

____

____

____

____

Tell us about <u>one</u> of the blocks that has a square face:

- the color of the block
- the size of the block
- the number of faces
- the number of edges
- the number of vertices

____

____

____

____

Student Mathematician: ____________________ Date: ____________

Organizing Ideas

Use the table below to record information about the blocks you think have a square face.

Block Name	Size and Color	Number of Faces	Number of Edges	Number of Vertices

Student Mathematician: Date:

Composing Rectangular Prisms

Materials

2 sets of tangram blocks

Directions

1. You and your partner, along with another pair, need to make different rectangular prisms.

   a. Remember that a rectangular prism has six rectangular faces.

   b. You will need to stack the blocks on top of each other to build the prism.

2. After you and your partner build one, show it to the other students in your group to make sure it is different from theirs.

3. Now try to build another different rectangular prism!

Example

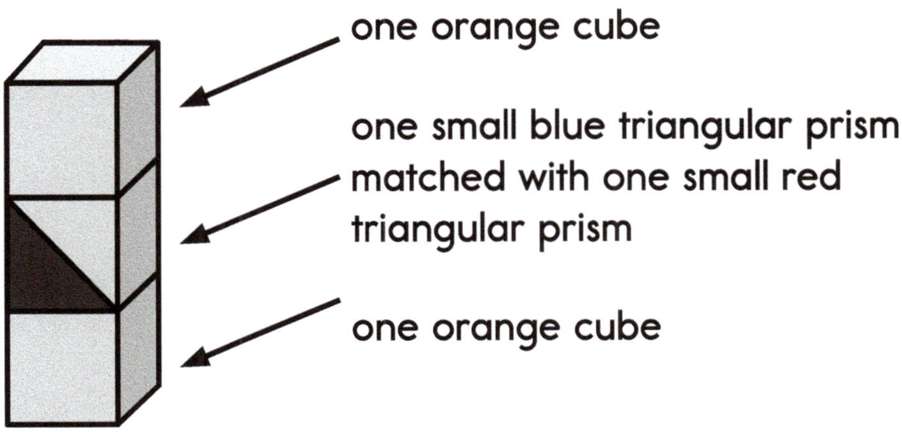

Student Mathematician: Date:

Composing Triangular Prisms

Materials

2 sets of tangram blocks

Directions

1. You and your partner, along with another pair, need to make different triangular prisms.
   - a. Remember that a triangular prism has two triangular faces and three rectangular faces.
   - b. You will need to stack the blocks on top of each other to make the prism.
2. After you and your partner build one, show it to the other students in your group to make sure it is different from theirs.
3. Now try to build another different triangular prism!

Example

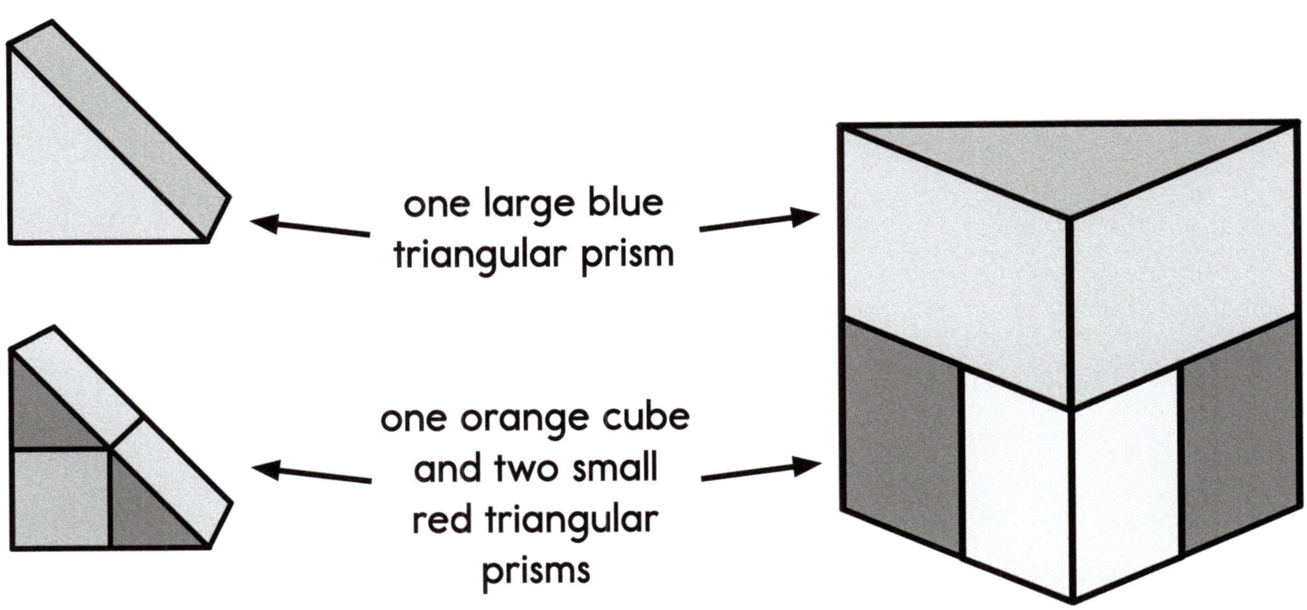

Ideas Needed

Shape Gallery at SMILE

The **S**ensational **M**ath **I**nstitute for **L**earning and **E**xploration

Needed:

Scale Models of Your Playground Climbing Equipment

Send us your designs!

Guidelines

Use your blocks to make a scale model of your playground climbing equipment.

1. Use 8–14 blocks from your two sets of blocks.
2. The blocks should cover the four faces on your floor design exactly.
3. Use at least two blocks for each piece of playground climbing equipment.
4. When two blocks are combined:
   a. their colors must be different.
   b. their faces must match exactly.

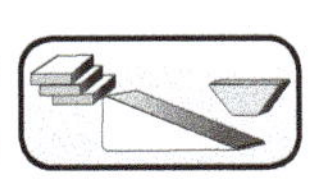

Student Mathematician: Date:

Equipment Design Checklist

Answer each question. After you have answered all the questions "yes," ask another pair to check your questions and respond "yes" or "no." If you disagree, talk about it and decide who is correct. Correct your design if necessary.

Question	You		Other Pair	
1. Did you use 8 to 14 blocks?	Yes	No	Yes	No
2. Do your blocks match your floor design?	Yes	No	Yes	No
3. Does each piece of equipment have at least two blocks?	Yes	No	Yes	No
4. When two blocks are combined, are their colors different?	Yes	No	Yes	No
5. When two blocks are combined, do their faces match exactly?	Yes	No	Yes	No

Student Mathematician: ______________________ Date: __________

THINK DEEPLY

Shape Gallery
at SMILE
The **S**ensational **M**ath **I**nstitute for **L**earning and **E**xploration

Choose one of your equipment pieces for us to use in the Shape Gallery. Help us understand your design!

a. Name it.

b. What shapes did you use to build your equipment piece?

c. How did you put the blocks together?

Final Ideas

Shape Gallery at SMILE

The Sensational Math Institute for Learning and Exploration

Needed:

2-Dimensional Scale Drawings of Your Playground Climbing Equipment

Send us your designs!

Guidelines

Help us get a better idea of your playground climbing equipment!

1. Pick one equipment piece.
2. Draw:
   - **a.** the front view
   - **b.** the back view
   - **c.** the left view
   - **d.** the right view
   - **e.** the top view
3. Color the shapes to match the blocks.

Student Mathematician: _______________ Date: _______________

Different Views

Front View

Back View

Left Side View

Right Side View

Top View

Notes to Dru and Teller:

Student Mathematician: ______________________________ Date: ______________

THINK DEEPLY

Shape Gallery
at SMILE
The **S**ensational **M**ath **I**nstitute for **L**earning and **E**xploration

We want to make sure we understand your design. Tell how one of the views matches your playground equipment piece.

Student Mathematician's Glossary

____________________	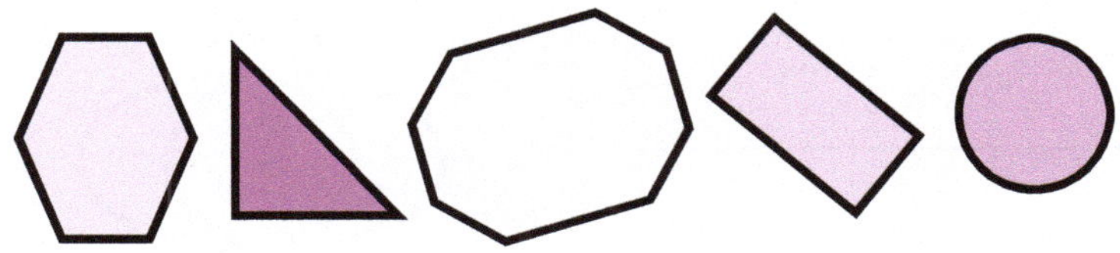
____________________	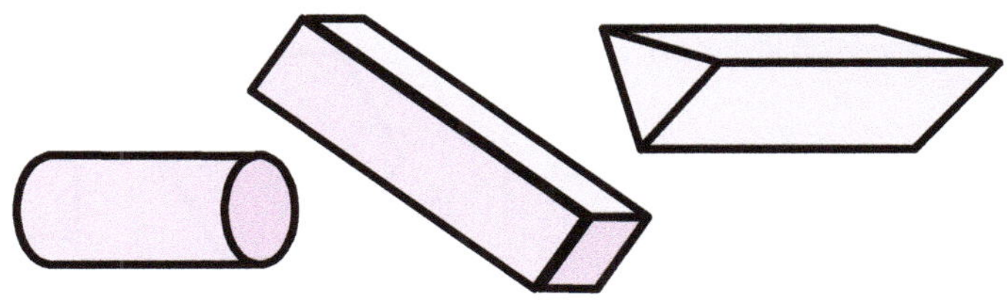
____________________	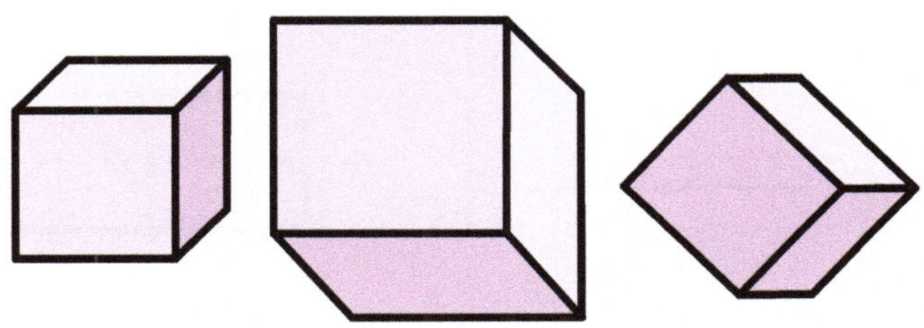
____________________	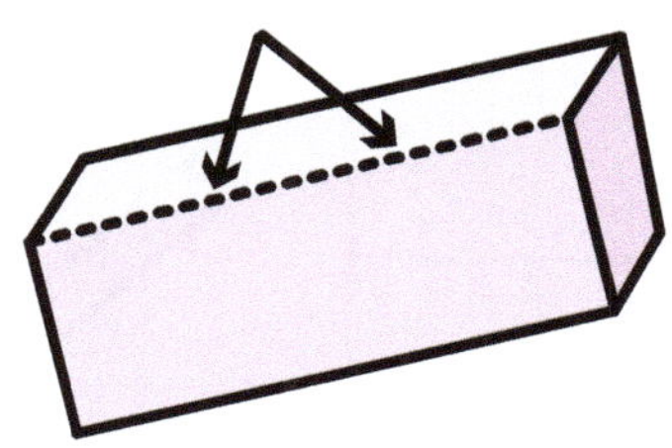
____________________	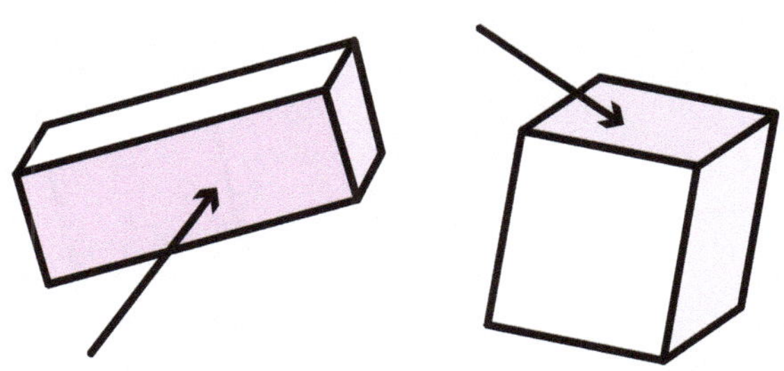

Student Mathematician's Glossary

________________	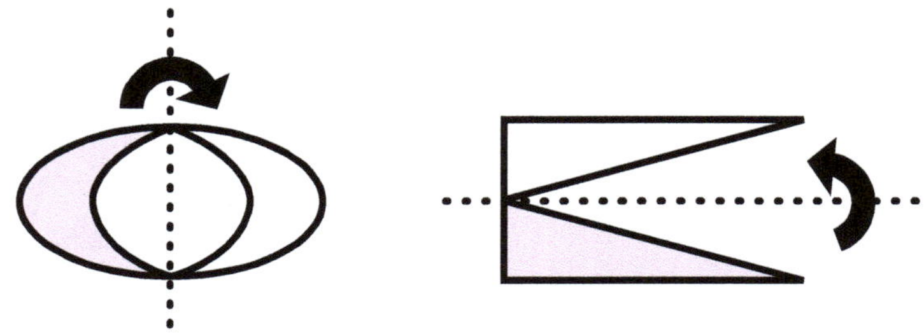
________________	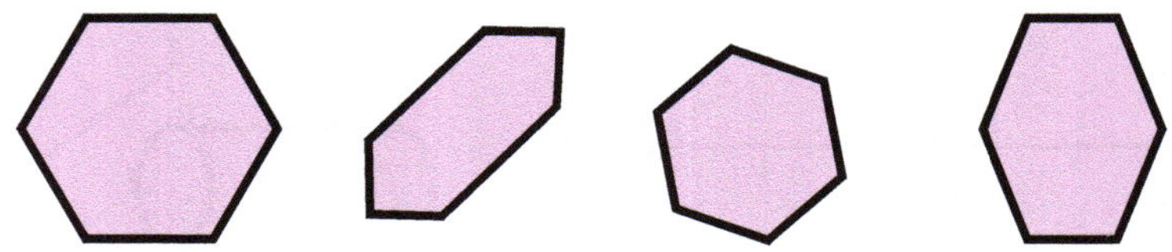
________________	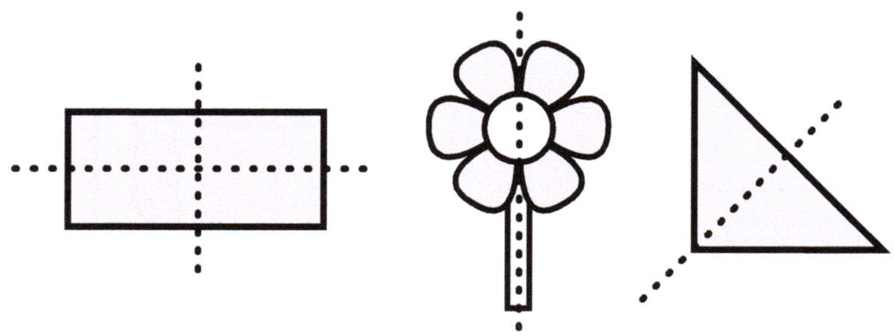
________________	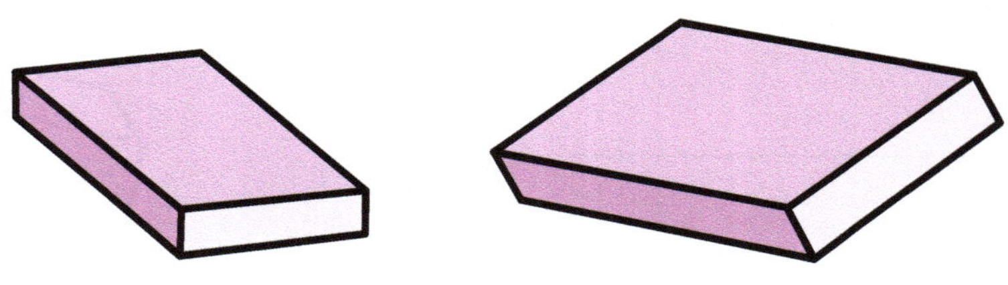
________________	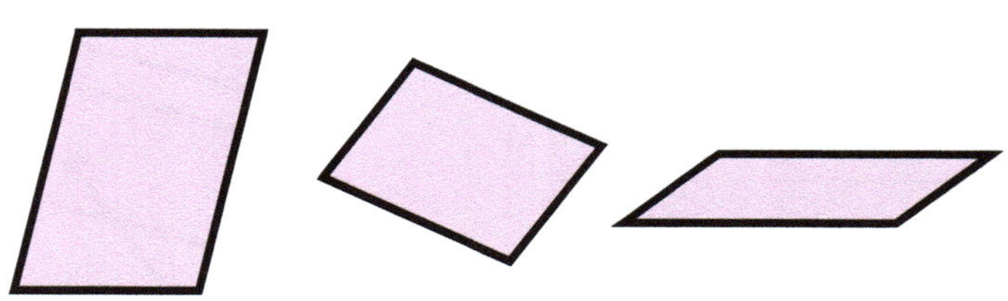

Student Mathematician's Glossary

____________________	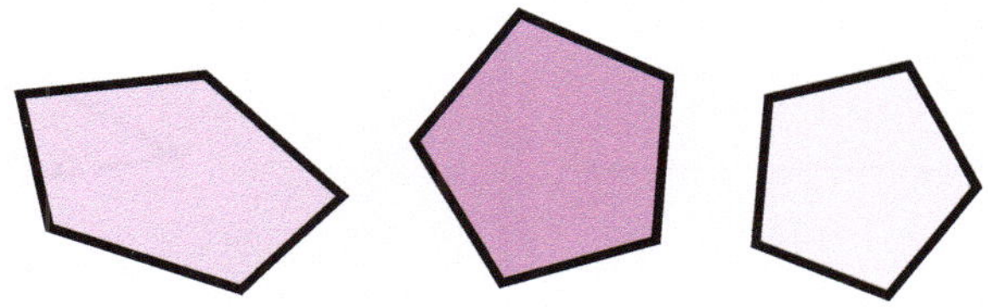
____________________	
____________________	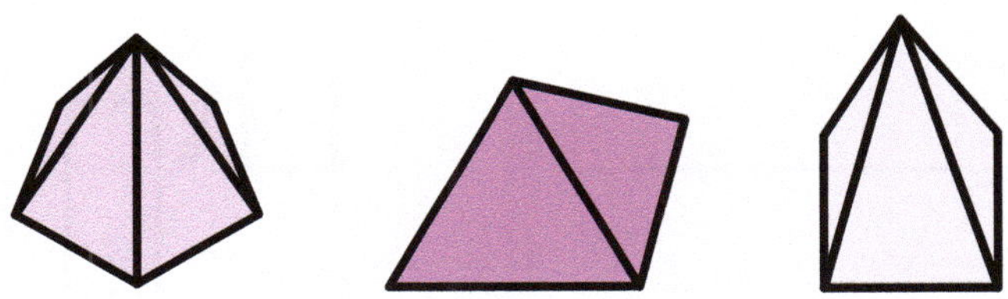
____________________	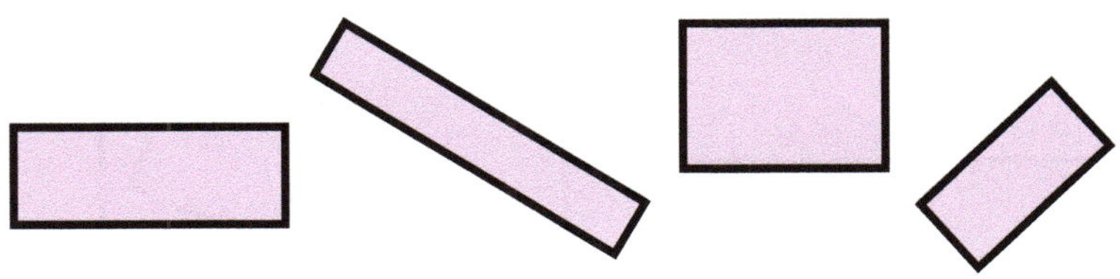
____________________	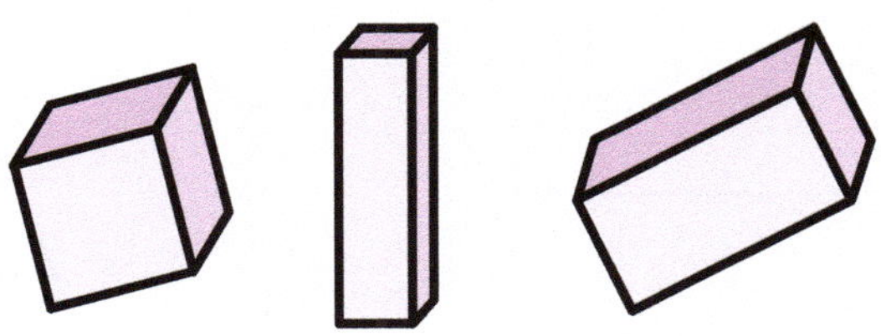

Student Mathematician's Glossary

	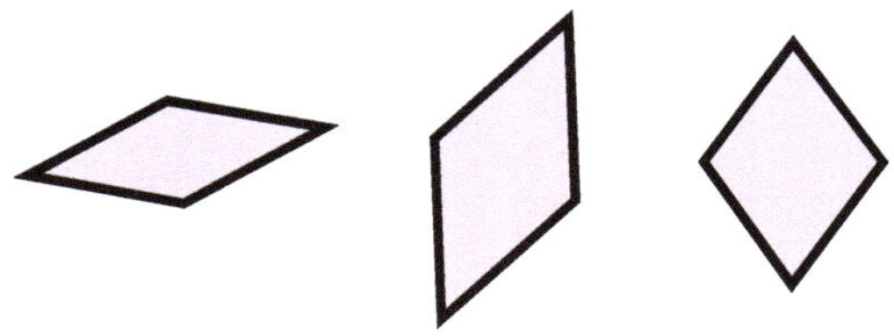
	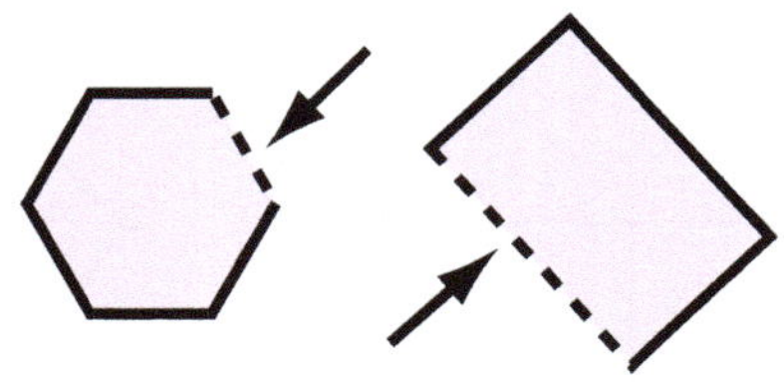
	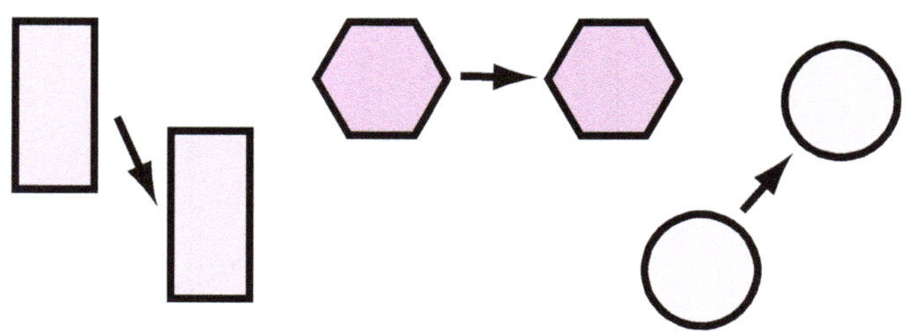
	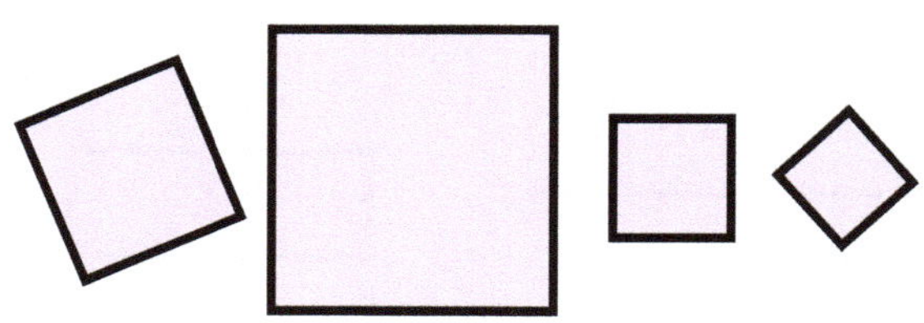
	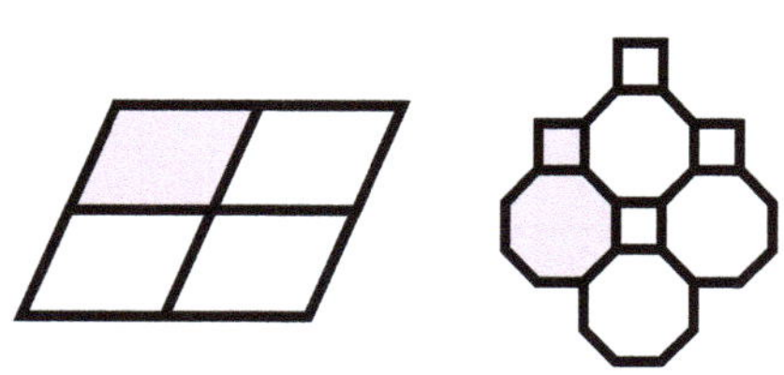

Student Mathematician's Glossary

______________________	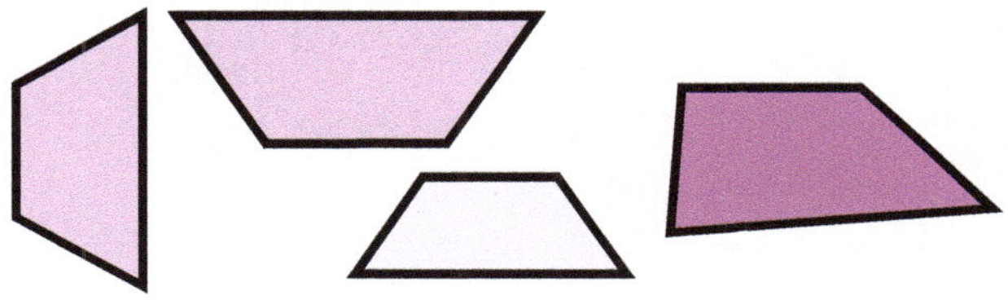
______________________	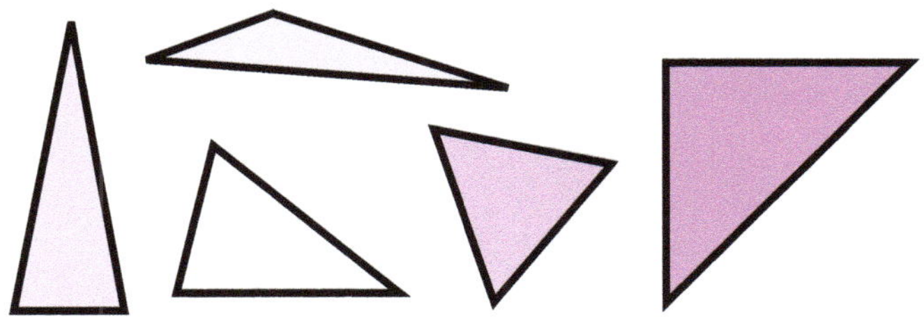
______________________	
______________________	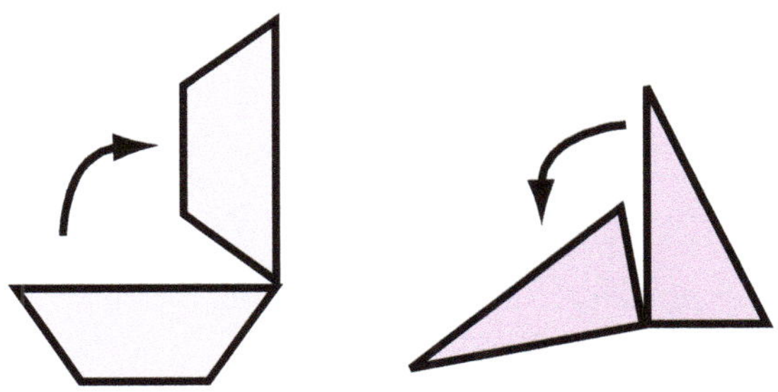
______________________	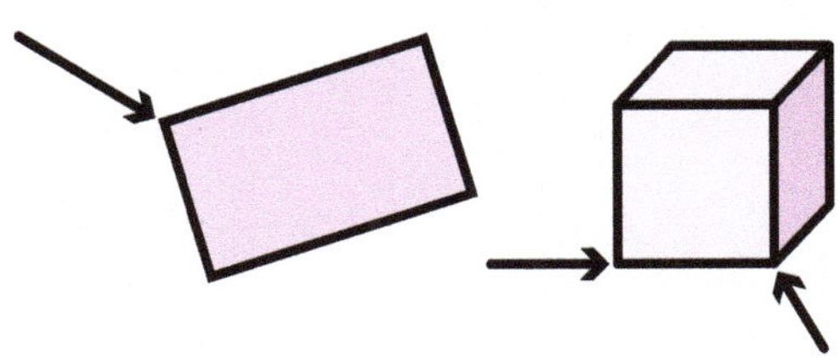

www.ingramcontent.com/pod-product-compliance
Ingram Content Group UK Ltd.
Pitfield, Milton Keynes, MK11 3LW, UK
UKHW050728071026
15590UKWH00012B/2318

9 781524 912130